P9-DBS-659

Space Travel

by
Jenny Tesar

Heinemann Interactive Library
Des Plaines, Illinois

Published by Heinemann Interactive Library,
an imprint of Reed Educational & Professional Publishing
1350 East Touhy Avenue, Suite 240 West
Des Plaines, IL 60018

01 00 99 98

10 9 8 7 6 5 4 3 2 1

ISBN 1-57572-581-9

Library of Congress Cataloging-in-Publication Data

Tesar, Jenny E.
Space travel / by Jenny Tesar.
p. cm. — (Space observer)
Includes bibliographical references and index.
Summary: Provides an introduction to space travel, describing what astronauts do, how they live and work in space, space shuttles and space stations, moon landings, and more.
ISBN 1-57572-581-9 (lib. bdg.)
1. Space flight—Juvenile literature. 2. Outer space—Exploration—Juvenile literature. [1. Manned space flight.]
I. Title. II. Series: Tesar, Jenny E. Space observer.
TL793.T425 1997
629.45—DC21 97-25178
CIP
AC

Acknowledgments
The author and publishers are grateful to the following for permission to reproduce copyright photographs:
Pages 4-5: ©Blackbirch Press, Inc.; pages 6, 18, 22-23: ©NASA/Science Source/Photo Researchers, Inc.; page 7: A.S.P./Science Source/Photo Researchers, Inc.; pages 8-9,
14: ©Julian Baum/Science Photo Library/Photo Researchers, Inc.; page 10: ©John Foster/Science Source/Photo Researchers, Inc.; page 11: Gazelle Technologies, Inc.; page 12: U.S. Geological Survey/Science Photo Library/Photo Researchers, Inc.;
pages 13, 16, 19: ©NASA; page 15: A. Gragera, Latin Stock/Science Photo Library/Photo Researchers, Inc.; page 17: ©NASA/Peter Arnold, Inc.; pages 20,
21: ©W. Kaufmann/JPL/SS/Photo Researchers, Inc.

Cover photograph: NASA/Peter Arnold, Inc.

Every effort has been made to contact copyright holders of any material reproduced in this book. Any omissions will be rectified in subsequent printings if notice is given to the publisher.

Some words are shown in bold, **like this**. You can find out what they mean by looking in the glossary.

Printed by Times Offset (M) Sdn. Bhd.

Contents

Going into Space

For thousands of years, people have wondered about space. They have dreamed of traveling to the stars and planets they could see in the sky.

The first person to travel into space was the Russian Yuri Gagarin. In 1961, he spent less than two hours in space. Since then, hundreds of people have gone into space. Some have traveled to the Moon.

Russian ***astronaut*** *Yuri Gagarin.*

Mission Control

A trip into space is called a mission. People on the ground control the mission. They use **radio signals** and computers to stay in touch with the spaceship.

This mission control is at the Kennedy Space Center.

Astronauts get their training at mission control.

People at mission control also train **astronauts.** The astronauts learn how to wear space suits, move about and work in space, and operate their spaceship.

Leaving Earth

Spaceships have to be very powerful to escape Earth's **atmosphere.** They get their energy from big, powerful rockets. A rocket is a kind of engine.

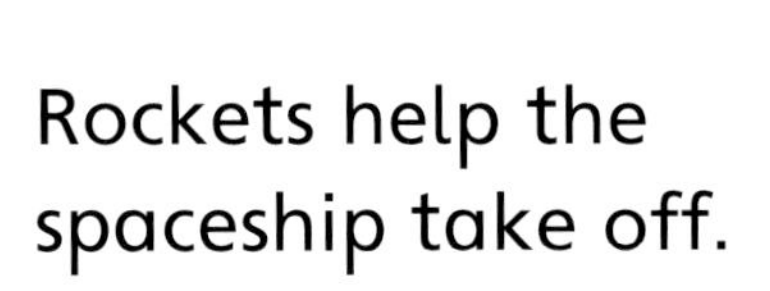

Rockets help the spaceship take off.

A spaceship heads for space.

Spaceships are attached to rockets so they can be **launched.** When the rocket fuel is lit, the rocket and spaceship shoot up into space.

Space Shuttle

A space **shuttle** is a small spaceship that travels back and forth between Earth and space.

Space shuttles carry **astronauts** and materials into space. The astronauts do tests to learn how space affects their bodies. They fix damaged **satellites.** They also study Earth and other objects in space.

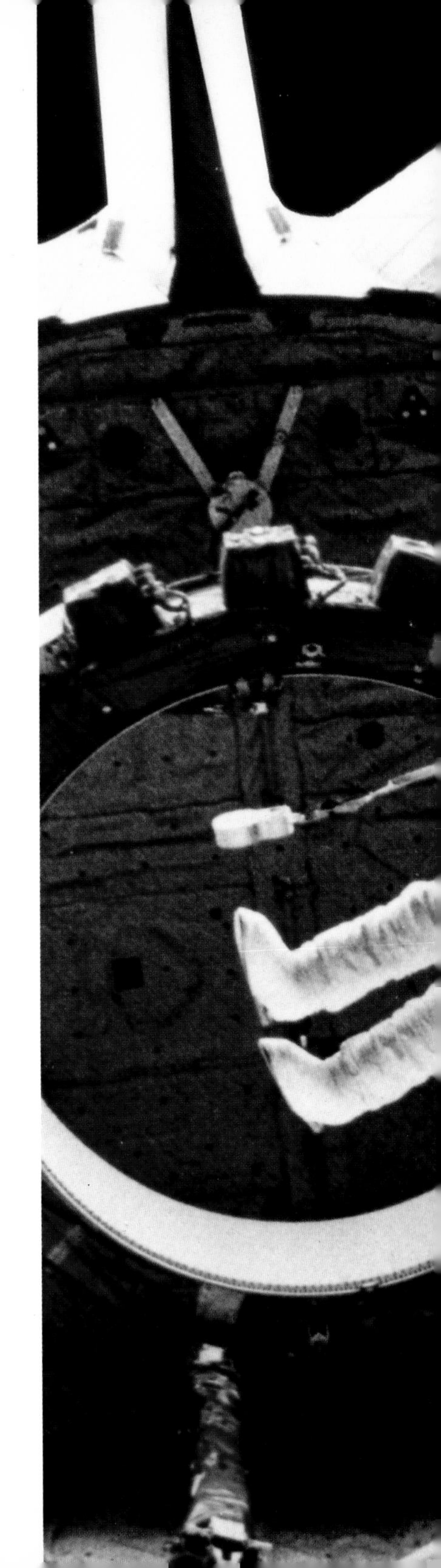

Two astronauts work in space while attached to their shuttle.

Living in Space

Space is very different from Earth. In space there is no air to breathe. There is no water or food. Spaceships that carry **astronauts** also must carry air, water, food, and other things that people need to live.

An astronaut in space talks with mission control.

Space suits protect astronauts against heat and cold.

Spaceships and space suits protect astronauts against the very hot and cold temperatures in space.

Working in Space

Inside their spaceship, **astronauts** wear light, comfortable clothes. Sometimes, astronauts must go outside their spaceship to work. Outside, they need to wear a special suit made of many layers of strong material.

Astronauts wear regular clothing when working inside.

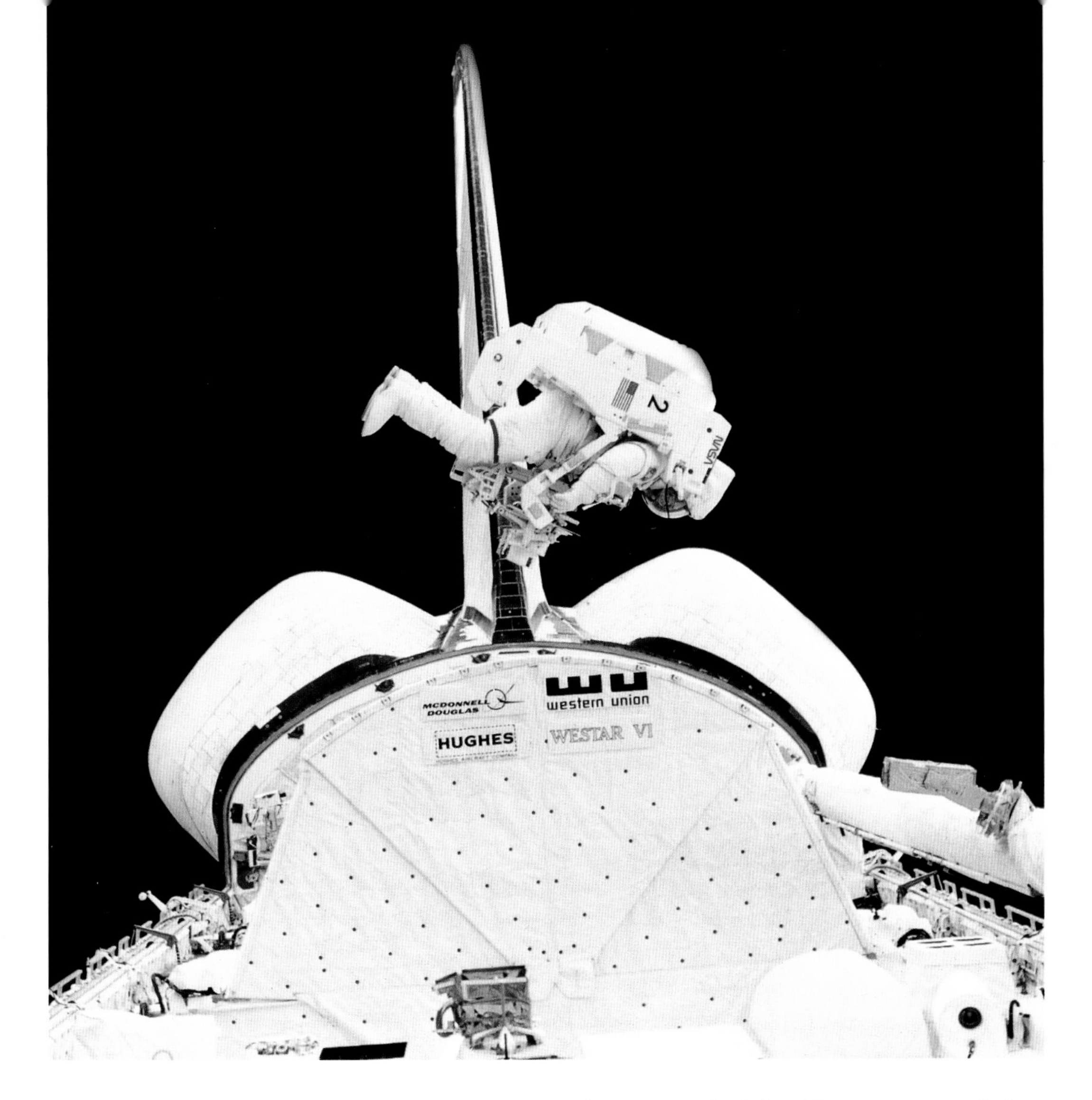

Astronauts wear space suits when outside the spaceship.

The suit contains air to breathe. The helmet has earphones and a microphone so the astronauts can talk to each other and to mission control.

Space Stations

A space station is a kind of **satellite.** It is a place where **astronauts** can live in space. Spaceships bring astronauts and supplies, such as food and water, to a space station.

In 1996, American Shannon Lucid lived in the Russian space station *Mir* for more than six months. She traveled 75 million miles and circled Earth 3,008 times!

Shannon Lucid works with another astronaut at the *Mir* space station.

Moon Landings

U.S. **astronauts** Neil Armstrong and Buzz Aldrin were the first people to walk on the Moon. They landed their **spacecraft,** *Apollo 11,* on the Moon on July 20, 1969.

Buzz Aldrin was one of the first astronauts to land on the moon.

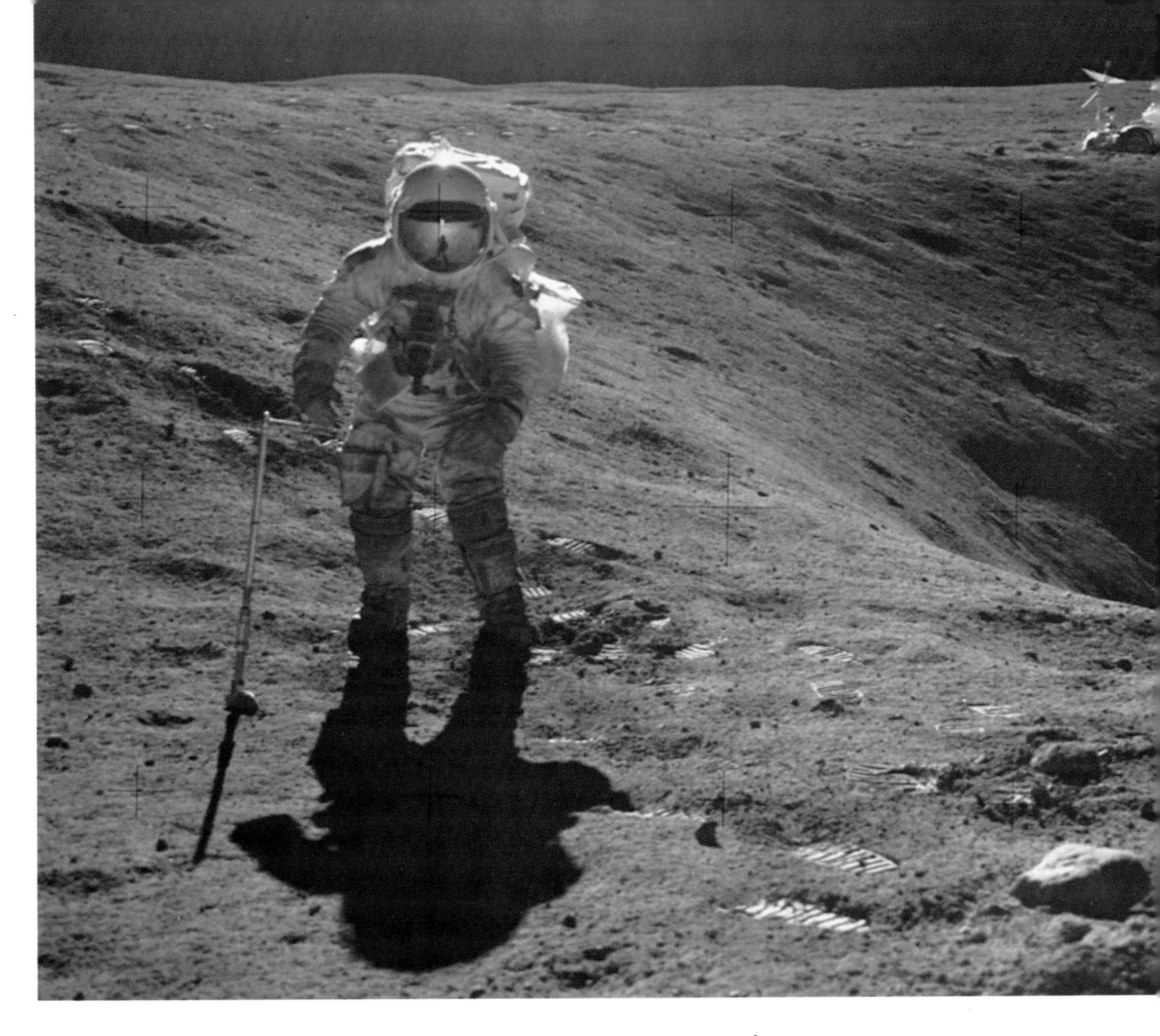

Astronauts have collected many moon rocks.

After *Apollo 11,* there were five other Moon landings. On most of these missions, astronauts collected hundreds of pounds of rock samples.

Space Probes

Space probes are **unmanned spacecraft** that explore outer space. Probes carry cameras, **radar**, and other instruments. They send information back to Earth.

Space probes have taken photos that scientists can use to make maps of space. In 1997, a space probe landed on Mars. It may tell us if there is life there.

An artist's picture of the space probe *Voyager 2* near Neptune and one of its moons.

Space Colonies

Someday, people may live in space **colonies** that have homes, schools, and farms. The colonies will be under big domes filled with air.

Maybe you will be one of these space people. Will you work in a factory on a giant space station? Take a vacation on the Moon? The story of space travel is only just beginning!

This imaginary space base could grow to be a space colony!

SPACE
48

Glossary

astronauts People who travel to space.

atmosphere Mixture of gases around a planet.

colonies Communities of people who have left their homeland to settle in a new place.

launched Sent into space.

radar Equipment that uses radio waves to find solid objects.

radio signals A form of energy.

satellite A **spacecraft** that travels in a path in space.

shuttle Something that moves back and forth from one place to another.

spacecraft A vehicle that travels to space.

unmanned Without people.

More Books to Read

Asimov, Isaac. *Exploring Outer Space: Rockets, Probes, and Satellites.* Milwaukee, WI: Gareth Stevens, 1995.

Donnelly, Judy. *Moonwalk: The First Trip to the Moon.* New York: Random Books for Young Readers, 1989.

Mayes, S. *What's Out in Space?* Tulsa, OK: Usborne, 1990

Maynard, Chris. *I Wonder Why Stars Twinkle and Other Questions about Space.* Las Vegas, NV: Kingfisher Books, 1993.

Index